NUREG-0696

I0475772

Functional Criteria for Emergency Response Facilities

Final Report

Manuscript Completed: December 1980
Date Published: February 1981

Division of Emergency Preparedness
Office of Inspection and Enforcement
U.S. Nuclear Regulatory Commission
Washington, D.C. 20555

ABSTRACT

This report describes the facilities and systems to be used by nuclear power plant licensees to improve responses to emergency situations. The facilities include the technical support center (TSC), onsite operational support center (OSC), and nearsite emergency operations facility (EOF), as well as a brief discussion of the emergency response function of the control room. The data systems described are the safety parameter display system (SPDS) and nuclear data link (NDL). Together, these facilities and systems make up the total emergency response facilities (ERFs). Licensees should follow the guidance provided both in this report and in NUREG-0654 (FEMA-REP-1), Revision 1 for design and implementation of the ERFs.

This document is being issued to establish criteria that the NRC staff intends to use in evaluating whether an applicant/licensee meets the requirements of 10 CFR 50, Appendix E, Article IV.E.8 and Appendix A, GDC 19. It is not a substitute for the regulations, and compliance is not a requirement. However, the use of criteria different from those set forth herein will be accepted only if the substitute criteria provide a basis for determining that the above-cited regulatory requirements have been met.

Contents

Contents (continued)

FIGURES

Contents (continued)

TABLES

FUNCTIONAL CRITERIA FOR EMERGENCY RESPONSE FACILITIES

1. INTRODUCTION

The facilities and systems described in this report are to be used by nuclear power plant licensees to provide improved emergency response to accidents. The facilities include the control room (CR), onsite technical support center (TSC), onsite operational support center (OSC), and nearsite emergency operations facility (EOF). The systems include the safety parameter display system (SPDS) and nuclear data link (NDL). Although the SPDS and the NDL are not "facilities," for ease of description these systems and the facilities (TSC, OSC, and EOF) will be referred to as the emergency response facilities (ERFs). The TSC, EOF, and CR are required facilities specified by NRC regulations, 10 CFR 50, Appendix E, Article IV.E.8 and Appendix A, GDC 19. Although not specifi-cally required by the regulations, in order to perform NRC's statutory responsi-bility to protect the public health and safety in the event of a radiological incident at a nuclear power plant and as a result of lessons learned from the Three Mile Island accident, the SPDS and the NDL system are included to provide assurance that licensees and offsite authorities will have adequate information to perform their tasks in an optimum manner. The OSC also is not specifically required by the regulations, but is needed to ensure that an adequate facility is provided for onsite emergency maintenance and for other support personnel to gather as a ready resource to support actions initiated by the CR. Thus, the SPDS, NDL, and OSC are integrally related to the emergency response facilities required by the regulations, and are needed to ensure that their functions can be effectively performed during an emergency. This document provides guidance on the functional criteria for the ERFs and on the integrated support these facilities will provide to the control room. Design and implementation of the ERFs should also be integrated with the licensee's implementation of NUREG-0654 (FEMA-REP-1), Revision 1, "Criteria for Preparation and Evaluation of Radio-logical Emergency Response Plans and Preparedness in Support of Nuclear Power Plants."* The report also provides detailed criteria for emergency response and certain emergency planning elements necessary for designing these facilities.

1.1 Background

The accident at Three Mile Island led to studies performed both within and outside NRC that identified the need for extensive improvements in the response of management to accidents at nuclear power plants. Some of the identified improvements include:

- Establishing formal licensee, local, State, and Federal organizations to better manage and effectively coordinate emergency response support;

- Developing integrated emergency response facilities and data systems to aid in this management;

- Providing for better information needed to assess conditions at a plant and its environs prior to, during, and following an accident;

---

*For complete information on material referenced in this report, refer to the Bibliography (Section 10) at the end of the text.

Providing an improved capability by licensee and Federal organizations to provide recommendations to State and local authorities on actions protecting the public; and

Providing transmission of more accurate information to Federal, State, and local emergency response organizations, and to the general public.

With respect to licensee activities and responsibilities, NRC has determined that the emergency response facilities and systems that will provide the necessary improvements are the TSC, OSC, EOF, SPDS and NDL. These facilities and systems will operate as an integrated system to support the control room in the mitigation of the consequences of accidents and to enhance the licensee's capability to respond to abnormal plant conditions. These facilities and systems will help to provide a graduated response capability that is dependent on the severity of an emergency. Severity conditions are classified into emergency classes (in order of increasing severity) by NUREG-0654, Revision 1, Appendix 1 as follows:

(1) Notification of Unusual Event
(2) Alert
(3) Site Area Emergency
(4) General Emergency

1.2 Control Room

The control room is the onsite location from which the nuclear power plant is operated. It contains the instrumentation, controls, and displays for:

- Nuclear systems,
- Reactor coolant systems,
- Steam systems,
- Electrical systems,
- Safety systems (including engineered safety features), and
- Accident monitoring systems.

The control room is staffed by licensed reactor operators, the senior reactor operators, and a senior reactor operator designated as the shift supervisor. Safe operation of the reactor and plant manipulations remain under the control of a licensed senior reactor operator, a reactor operator, or shift supervisor at all times.

During abnormal operating conditions, the complexity of licensee responsibilities increases significantly. These responsibilities include actions taken to:

- Diagnose the abnormal conditions;
- Perform corrective actions;
- Mitigate the abnormal conditions;
- Manage plant operations;
- Manage emergency response;
- Inform Federal, State, and local officials;
- Recommend public protective measures to State and local officials;
- Restore the plant to a safe condition; and
- Recover from the abnormal conditions.

Initially, the control room personnel must assume all of these responsibilities.

1.3 Emergency Response Facilities

In helping personnel in the control room to mitigate the consequences of accidents and respond to abnormal operating conditions, the emergency response facilities (ERFs) shall function during emergencies to provide the following services:

- Help the reactor operators determine the plant safety status;

- Relieve the reactor operators of peripheral duties and communications not directly related to reactor system manipulations;

- Prevent congestion in the control room;

- Provide assistance to the operators by technical personnel who have comprehensive plant data at their disposal;

- Provide a coordinated emergency response by both technical and management personnel;

- Provide reliable communications between onsite and offsite emergency response personnel;

- Provide a focal point for development of recommendations for offsite actions; and

- Provide relevant plant data to the NRC for its analysis of abnormal plant operating conditions.

Many of the ERF functions will be performed through the use of one or more data systems that will gather, store, and process data for display in the control room, TSC, EOF, and NRC Operations Center. A partially or fully integrated ERF data system or separate data systems may be used to fulfill the ERF functions. A recent-generation plant process computer is one data system that could be used to gather and process these data for display in the ERFs. However, use of a process computer requires that it meet all of the criteria prescribed later in this document. If it cannot meet these criteria, independent data systems must be installed.

The personnel designated for duty in the ERFs shall be trained to follow procedures specified in the licensee's emergency plan to ensure timely emergency response. Detailed management plans, facility staffing, and task assignments of these emergency response personnel are site specific and shall be part of the licensee's emergency plan developed from the guidance provided by NUREG-0654, Revision 1.

1.3.1 Technical Support Center

The technical support center (TSC) is an onsite facility located close to the control room that shall provide plant management and technical support to the reactor operating personnel located in the control room during emergency conditions. It shall have technical data displays and plant records available to assist in the detailed analysis and diagnosis of abnormal plant conditions and any significant release of radioactivity to the environment. The TSC shall be the primary communications center for the plant during an emergency. A senior official, designated by the licensee, shall use the resources of the TSC to assist the control room operators by handling the administrative items, technical evaluations, and contact with offsite activities, relieving them of these functions. The TSC facilities may also be used for performing normal functions, such as shift technical supervisor and plant operations/maintenance analysis functions, as well as for emergencies.

1.3.2 Operational Support Center

The operational support center (OSC) is an onsite assembly area separate from the control room and the TSC where licensee operations support personnel shall report in an emergency. There shall be direct communications between the OSC and the control room and between the OSC and the TSC so that the personnel reporting to the OSC can be assigned to duties in support of emergency operations.

1.3.3 Emergency Operations Facility

The emergency operations facility (EOF) is a nearsite support facility for the management of overall licensee emergency response (including coordination with Federal, State, and local officials), coordination of radiological and environmental assessments, .and determination of recommended public protective actions. The EOF shall have appropriate technical data displays and plant records to assist in the diagnosis of plant conditions to evaluate the potential or actual release of radioactive materials to the environment. A senior licensee official in the EOF shall organize and manage licensee offsite resources to support the TSC and the control room operators.

1.3.4 Safety Parameter Display System

The safety parameter display system (SPDS) provides a display of plant parameters from which the safety status of operation may be assessed in the control room, TSC, and EOF. The primary function of the SPDS is to help operating personnel in the control room make quick assessments of plant safety status. Duplication of the SPDS displays in the TSC and EOF will improve the exchange of information between these facilities and the control room and assist corporate and plant management in the decision-making process. The SPDS shall be operated during normal operations and during all classes of emergencies. The SPDS should have the flexibility to allow future modifications to be incorporated, such as the capability to handle operator interaction and diagnostic analysis.

1.3.5 Nuclear Data Link

The nuclear data link (NDL), a data transmission system, will be designed to send a set of variables from the plant to the NRC Operations Center. These data will be used for analyses by the NRC headquarters technical support groups and NRC Executive Team. The NDL will transmit information that will aid NRC in its role of providing advice and support to the nuclear power plant licensee, State and local authorities, and other Federal officials.

1.4 Activation and Use

The activation and use of the ERFs shall be determined by the emergency class and by the specific conditions surrounding an accident and shall be specified in the licensee's emergency plan. As a minimum, within the unavailability criteria, the following conditions shall apply:

(1) The SPDS shall be operational during all plant operating conditions, including accidents.

(2) Activation of the onsite TSC and OSC is optional for a Notification of Unusual Event emergency class, and is required for Alert and higher classes.

(3) Activation of the nearsite EOF is optional for Notification of Unusual Event and Alert emergency classes, and is required for Site Area Emergency and General Emergency classes.

(4) Data shall be available for transmittal through the NDL during all plant operating conditions above cold shutdown.

Until the TSC, OSC, and EOF are activated, all functions of these facilities shall be performed in the control room. When the TSC is functional, emergency response functions, except direct supervision of reactor operations and manipulation of reactor system controls, shall shift to the TSC. Plant administration, technical support functions, and contact with offsite activities to assist the control room operators shall be performed in the TSC throughout the course of an accident. The OSC shall provide a place for operations support personnel to be in direct communication with the control room and other operations managers for assignment to duties in support of emergency operations. When the EOF is activated, the functions of providing overall emergency response management, monitoring and assessing radiological effluent and the environs, making offsite dose projections, providing recommendations to State and local officials, and coordinating with Federal officials will shift to the EOF. See Table 1 for an outline of the transfer of emergency response functions from the control room to the TSC and EOF under the various emergency classes.

The level of staffing of the ERFs may vary according to the severity of the emergency condition. The staffing criteria for each emergency class shall be fully detailed in the licensee's emergency plan.

1.5 Reliability

The data systems, instrumentation, and facilities of the ERFs shall be designed and constructed to provide a very high degree of reliability. The reliability criteria for the ERF systems, instrumentation, and facilities shall be described in terms of unavailability. This unavailability is specified in terms of an operational unavailability goal of 0.01 that is applicable to all ERFs when the reactor is above cold shutdown status. This operational unavailability goal shall be defined in units of time as:

$$\text{Operational unavailability} = \frac{\text{Downtime}}{\text{Operating time}}$$

Table 1. Transfer of Emergency Response
Functions from the Control Room to the
Technical Support Center and the Emergency Operations Facility

Emergency Response Functions	Emergency Class			
	Notification of Unusual Event	Alert	Site Area Emergency	General Emergency
Supervision of reactor operations and manipulation of controls	CR	CR	CR	CR
Management of plant operations	CR(TSC)	TSC	TSC	TSC
Technical support to reactor operations	CR(TSC)	TSC	TSC	TSC
Management of corporate emergency response resources	CR(TSC,EOF)	TSC(EOF)	EOF	EOF
Radiological effluent and environs monitoring, assessment, and dose projections	CR(TSC,EOF)	TSC(EOF)	EOF	EOF
Inform Federal, State, and local emergency response organizations and make recommendations for public protective actions	CR(TSC,EOF)	TSC(EOF)	EOF	EOF
Event monitoring by NRC regional emergency response team*	(CR)	TSC(EOF)	TSC&EOF	TSC&EOF
Management of recovery operations	CR(TSC,EOF)	TSC(EOF)	EOF	EOF
Technical support of recovery operations	CR(TSC,EOF)	TSC	TSC	TSC

Note: (CR), (TSC), (EOF), or (TSC, EOF) indicates that activation of this facility (or the performance of this function) is optional for the indicated emergency class.

*One NRC individual also may be stationed in the control room.

where:

> . Downtime = any length of time the data systems, instrumentation, or facilities are unavailable when the reactor is above cold shutdown status because of the following:
>
> > • inability to perform its intended function;
> >
> > • impaired ability to perform its intended function due to degraded circuits, equipment, power supplies, or instrumentation (this shall not include truly redundant equipment such as core thermocouples or computer peripherals);
> >
> > • unreliable performance due to the lack of adequate sensor data; and
> >
> > • scheduled outages to perform preventive maintenance on instrumentation, equipment, power supplies, or sensors. (The design of the systems and facilities shall be to limit these scheduled outages to no more than 16 hours per calendar quarter, and ERFs must be capable of becoming fully operational within 30 minutes during these outages.)

Operating time = any length of time the reactor is above cold shutdown status.

In addition to the operational criterion during reactor operation, the SPDS also has an unavailability goal while the reactor is in cold shutdown status. For cold shutdown status including the refueling mode, the specified unavailability goal of the SPDS is 0.2. This SPDS cold shutdown unavailability shall be defined in units of time as:

$$\text{SPDS cold shutdown unavailability} = \frac{\text{Downtime}}{\text{Cold shutdown time}}$$

where:

> Downtime = any length of time the SPDS data system, instrumentation, and power supplies are unavailable when the reactor is in cold shutdown status because of the following:
>
> > • inability to perform its intended function;
> >
> > • impaired ability to perform its intended function due to degraded circuits, instrumentation, or power supplies;
> >
> > • unreliable performance due to the lack of sensor data;
> >
> > • scheduled outages for preventive maintenance of instrumentation, power supplies, or sensors. (The

7

SPDS must be able to be made available within 30 minutes during such outages.)

Cold shutdown time = any length of time the reactor is in a cold shutdown status and refueling mode.

The TSC, EOF, and NDL have no cold shutdown unavailability goals while the reactor is in cold shutdown status.

2. TECHNICAL SUPPORT CENTER

2.1 Function

The onsite technical support center (TSC) will provide the following functions:

- Provide plant management and technical support to plant operations personnel during emergency conditions.

- Relieve the reactor operators of peripheral duties and communications not directly related to reactor system manipulations.

- Prevent congestion in the control room.

- Perform EOF functions for the Alert Emergency class and for the Site Area Emergency class and General Emergency class until the EOF is functional.

The TSC shall be the emergency operations work area for designated technical, engineering, and senior licensee management personnel; any other licensee-designated personnel required to provide the needed technical support; and a small staff of NRC personnel. The primary role of NRC in the TSC will be to observe plant operation and provide advisory support to plant management. A senior licensee official shall use the resources of the TSC to provide guidance and technical assistance to the operating supervisor in the control room. However, all manipulations shall be performed by the control room licensed operators.

The TSC shall have facilities to support the plant management and technical personnel who will be assigned there during an emergency and will be the primary onsite communications center for the plant during the emergency. TSC personnel shall use the TSC data system to analyze the plant steady-state and dynamic behavior prior to and throughout the course of an accident. The results of this analysis will be used to provide guidance to the control room operating personnel in the management of abnormal conditions and in accident mitigation. TSC personnel will also use the environmental and radiological information available from the TSC data system to perform the necessary functions of the EOF when this facility is not operational. The TSC also may be used to provide technical support during recovery operations following an emergency.

The TSC facilities may be used by designated operating personnel for normal daily operations as well as for training and emergency drills. Use of the TSC facility during normal operation shall be limited to activities that will not degrade TSC preparedness to react to abnormal conditions or reduce TSC systems reliability. The shift technical advisor (STA) may use the TSC facilities in performing his duties during normal operating conditions in addition to using them during emergencies.

Licensees who cannot meet the criteria for location, size, and habitability for the TSC must submit to NRC a request for an exception. This request must include justification for the exception and an alternate proposal. NRC will review requests for exceptions on a case-by-case basis.

2.2 Location

The onsite TSC is to provide facilities near the control room for detailed analyses of plant conditions during abnormal conditions or emergencies by trained and competent technical staff. During recent events at nuclear power plants, telephone communications between the facilities were ineffective in providing all of the necessary management interaction and technical information exchange. This demonstrates the need for face-to-face communications between TSC and control room personnel. To accomplish this, the TSC shall be as close as possible to the control room, preferably located within the same building. The walking time from the TSC to the control room shall not exceed 2 minutes. This close location will facilitate face-to-face interaction between control room personnel and the senior plant manager working in the TSC. This proximity also will provide access to information in the control room that is not available in the TSC data system.

Provisions shall be made for the safe and timely movement of personnel between the TSC and the control room under emergency conditions. These provisions shall include consideration of the effects of direct radiation and airborne radio-activity from inplant sources on personnel traveling between the two facilities. Anticontamination clothing, respiratory protection, and other protective gear may be used to help protect personnel in transit. The 2-minute travel time between the TSC and the control room does not include time required to put on any necessary radiological protective gear, but it does include the time required to clear any security checkpoints. There should be no major security barriers between these two facilities other than access control stations for the TSC and control room.

2.3 Staffing and Training

Upon activation of the TSC, designated personnel shall report directly to the TSC and achieve full functional operation within 30 minutes. Activation of the TSC shall ensure that only designated operating personnel are in the control room during the emergency and that needed technical support will be provided without obstructing actual plant manipulations or overcrowding the control room.

The licensee-designated TSC staff shall consist of sufficient technical, engineering, and senior designated licensee officials to provide the needed support to the control room during emergency conditions. Consultants also may be designated by the licensee to augment utility resources in the TSC. A senior designated licensee official shall coordinate activities in the TSC and interface with the control room, the OSC, and the EOF.

The level of staffing of the TSC may vary according to the severity of the emergency condition. The staffing for each emergency class shall be fully detailed in the licensee's emergency plan.

9

For the TSC to function effectively, TSC staff personnel must be aware of their responsibilities during an accident. The licensee shall, therefore, develop training programs for these personnel. In addition, to maintain proficiency, the TSC staff shall participate in TSC activation drills that shall be conducted periodically in accordance with the licensee's emergency plan. Operating procedures and staff training in the use of data systems and instrumentation shall contain guidance on the limitations of instrument readings including whether the information can be relied upon following such events as accidents resulting from earthquakes or the release of radiation.

2.4 Size

The TSC may be housed in a complex of directly adjacent areas. It shall be large enough to provide:

- Working space, without crowding, for the personnel assigned to the TSC at the maximum level of occupancy (minimum size of working space provided shall be approximately 75 sq ft/person);

- Space for the TSC data system equipment needed to acquire, process, and display data used in the TSC;

- Sufficient space to perform repair, maintenance, and service of equipment, displays, and instrumentation;

- Space for data transmission equipment needed to transmit data originating in the TSC to other locations;

- Space for personnel access to functional displays of TSC data;

- Space for unhindered access to communications equipment by all TSC personnel who need communications capabilities to perform their functions;

- Space for storage of and/or access to plant records and historical data; and

- A separate room adequate for at least three persons to be used for private NRC consultations.

The TSC working space shall be sized for a minimum of 25 persons, including 20 persons designated by the licensee and five NRC personnel. This minimum size shall be increased if the maximum staffing level specified by the licensee's emergency plan exceeds 20 persons.

2.5 Structure

The TSC complex must be able to withstand the most adverse conditions reasonably expected during the design life of the plant including adequate capabilities for (1) earthquakes, (2) high winds (other than tornadoes), and (3) floods.

The TSC need not meet seismic Category I criteria or be qualified as an engineered safety feature (ESF). Normally, a well-engineered structure will provide an adequate capability to withstand earthquakes. Winds and floods with a 100-year-recurrence frequency are acceptable as a design basis. Existing buildings may be used to house the TSC complex if they satisfy the above minimum criteria.

2.6 Habitability

Since the TSC is to provide direct management and technical support to the control room during an accident, it shall have the same radiological habitability as the control room under accident conditions. TSC personnel shall be protected from radiological hazards, including direct radiation and airborne radioactivity from inplant sources under accident conditions, to the same degree as control room personnel. Applicable criteria are specified in General Design Criterion 19; Standard Review Plan 6.4; and NUREG-0737, "Clarification of TMI Action Plan Requirements," Item II.B.2.

The TSC ventilation system shall function in a manner comparable to the control room ventilation system. The TSC ventilation system need not be seismic Category I qualified, redundant, instrumented in the control room, or automatically activated to fulfill its role. A TSC ventilation system that includes high-efficiency particulate air (HEPA) and charcoal filters is needed, as a minimum. Sufficient potassium iodide shall be provided for use by TSC and control room personnel. The capacity of the installed TSC ventilation filter system shall be independent of these thyroid-blocking provisions.

To ensure adequate radiological protection of TSC personnel, radiation monitoring systems shall be provided in the TSC. These monitoring systems may be composed of installed monitors or portable monitoring equipment dedicated to the TSC. These systems shall continuously indicate radiation dose rates and airborne radioactivity concentrations inside the TSC while it is in use during an emergency. These monitoring systems shall include local alarms with trip levels set to provide early warning to TSC personnel of adverse conditions that may affect the habitability of the TSC. Detectors shall be able to distinguish the presence or absence of radioiodines at concentrations as low as 10^{-7} microcuries/cc.

Equipment that protects personnel shall be provided in the TSC for the staff who must travel between the TSC and the control room or the EOF under adverse radiological conditions. Protective equipment also shall be provided to allow TSC personnel to continue to function during the presence of low-level airborne radioactivity or radioactive surface contamination. Anticontamination clothing and respiratory protective gear are examples of equipment that shall be provided. This equipment shall be properly maintained to assure availability during an emergency.

If the TSC becomes uninhabitable, the TSC plant management function shall be transferred to the control room.

2.7 Communications

The TSC will be the primary onsite communications center for the nuclear power plant during an emergency. It shall have reliable voice communications to the control room, the OSC, the EOF, and the NRC. The primary function of this voice communication system will be plant management communications and the immediate exchange of information on plant status and operations. Provisions for communications with State and local operations centers also shall be provided in the TSC to provide early notification and recommendations to offsite authorities prior to activation of the EOF.

The TSC voice communications facilities shall include means for reliable primary and backup communication. The TSC voice communications may include private telephones, commercial telephones, radio networks, and intercommunication systems as appropriate to accomplish the TSC functions during emergency operating conditions. Existing licensee communications systems may be used if the systems can be demonstrated to be reliable under emergency conditions and if they are adequate to meet the added TSC requirements. The licensee shall provide a means for TSC telephone access to commercial telephone common-carrier services that bypasses any onsite or local offsite telephone switching facilities that may be susceptible to loss of power during emergencies. The licensee shall ensure that spare commercial telephone lines to the plant are available for use by the TSC during emergencies.

The TSC voice communications equipment shall include:

- Hotline telephone (located in the NRC consultation room) on the NRC emergency notification system (ENS) to the NRC Operations Center;

- Dedicated telephone (located in the NRC consultation room) on the NRC health physics network (HPN);

- Dedicated telephones for management communications with direct access to the control room, the OSC, and the EOF;

- Dial telephones that provide access to onsite and offsite locations;

- Intercommunications systems between work areas of the TSC, if needed for the TSC functional performance or if the TSC is comprised of separate functional areas; and

- Communications to licensee mobile monitoring teams and to State and local operations centers prior to EOF activation.

The TSC communication system shall also include designated telephones (in addition to the ENS and HPN telephones) for use by NRC personnel. The licensee shall provide at least two dial telephone lines for such NRC use when the TSC is activated. The licensee shall also furnish the onsite access facilities and cables to the NRC for the ENS and HPN telephones.

Facsimile transmission capability between the TSC, the EOF, and the NRC Operations Center shall also be provided.

2.8 Instrumentation, Data System Equipment, and Power Supplies

Equipment shall be provided to gather, store, and display data needed in the TSC to analyze plant conditions. The data system equipment shall perform these functions independent of actions in the control room and without degrading or interfering with control room and plant functions. TSC instrumentation data system equipment and power supplies need not meet safety-grade or Class 1E requirements. When signals to the TSC are received from sensors providing signals to safety system equipment or displays, suitable isolation shall be provided in accordance with GDC 24 to ensure that the TSC systems will not degrade performance of the safety system equipment or displays.

The TSC electrical equipment load shall not degrade the capability or reliability of any safety-related power source. Circuit transients or power-supply failures and fluctuations shall not cause a loss of any stored data vital to the TSC functions. Sufficient alternate or backup power sources shall be provided to maintain continuity of TSC functions and to immediately resume data acquisition, storage, and display of TSC data if loss of the primary TSC power sources occurs.

The total TSC data system reliability shall be designed to achieve an operational unavailability goal of 0.01 during all plant operating conditions above cold shutdown as described in Section 1.5.

The plant data processor (plant process computer) may provide, if stringent conditions are met, the TSC data system functions or may have data for the TSC system transmitted through it. In either case, the performance of the plant data processor and its related instrumentation and equipment shall be included when determining the TSC data system unavailability. In addition, the computational capacity and data throughput of the plant processor must be sufficient to accommodate the combined computational and input-output loads of the TSC system and the other functions being performed by the plant processor. TSC data processed through the plant data processor shall be continuously available to the TSC with the same reliability and accuracy required of a data system designed to operate independently from the plant data processor. Use of the plant data processor shall not degrade the integrity of data supplied to the TSC or the security of the software used to process TSC data.

The SPDS display equipment used in the TSC need not be seismically qualified; it need only meet the TSC data system equipment reliability and performance criteria. The design of the TSC data system equipment shall incorporate human factors engineering with consideration for both operating and maintenance personnel.

2.9 Technical Data and Data System

The TSC technical data system shall receive, store, process, and display information acquired from different areas of the plant as needed to perform the TSC function. The data available for display in the TSC must enable the plant management, engineering, and technical personnel assigned there to aid the control room operators in handling emergency conditions. The data system shall provide access to accurate and reliable information sufficient to determine:

- Plant steady-state operating conditions prior to the accident,

- Transient conditions producing the initiating event, and

13

- Plant systems dynamic behavior throughout the course of the accident.

The TSC data system may be used for:

- Reviewing the accident sequence,

- Determining appropriate mitigating actions,

- Evaluating the extent of any damage, and

- Determining plant status during recovery operations.

The data set available to the TSC data system must be complete enough to permit accurate assessment of the accident without interference from the control room emergency operation. As a minimum, the set of Type A, B, C, D, and E variables specified in Regulatory Guide 1.97, Revision 2, "Instrumentation for Light-Water-Cooled Nuclear Power Plants to Assess Plant and Environs Conditions During and Following an Accident," shall be available for display and printout in the TSC. In addition, all sensor data and calculated variables not specified in Reg. Guide 1.97 but included in the data sets for the SPDS, for the EOF, or for transmission to offsite locations shall be available for display. The accuracy of the data displayed shall not be significantly less than the accuracy of comparable data displayed in the control room. The time resolution of data acquisition shall be sufficient to provide data without loss of information during transient conditions. The time resolution for each sensor signal will depend on the potential transient behavior of the variable being measured. The TSC data displays of Reg. Guide 1.97 variables shall meet the criteria for TSC data but need not meet the Reg. Guide 1.97 design and qualification criteria for display of those variables in the control room.

Data storage and recall capability shall be provided for the TSC data set. At least 2 hours of pre-event data and 12 hours of post-event data shall be recorded. The sample frequency shall be chosen to be consistent with the use of the data. Capacity to record at least two weeks of additional post-event data with reduced-time resolution shall be provided. Archival data storage and the capability to transfer data between active memory and archival data storage without interrupting TSC data acquisition and displays shall be provided for all TSC data.

A sufficient number of data display and printout devices shall be provided in the TSC to allow all TSC personnel to perform their assigned tasks with unhindered access to data. The TSC displays shall include, but not be limited to, alphanumeric and/or graphical representations of:

- Plant systems variables,
- In-plant radiological variables,
- Meteorological information, and
- Offsite radiological information.

Trend information display and time-history display capability is needed in the TSC to give the TSC personnel a dynamic view of the plant status during abnormal operating conditions. The TSC displays shall be designed so that callup, manipulation, and presentation of data can be easily performed. The TSC data display formats shall present information so that it can be easily understood by the TSC personnel performing analyses.

The SPDS shall also be displayed in the TSC. This duplication will improve the exchange of information between the control room and the TSC. If the SPDS system in the control room is composed of multiple display units, multiple displays must also be provided in the TSC.

2.10 Records Availability and Management

The TSC shall have a complete and up-to-date repository of plant records and procedures at the disposal of TSC personnel to aid in their technical analysis and evaluation of emergency conditions. In particular, up-to-date as-built drawings of the plant systems are needed to diagnose sensor data, evaluate data inconsistencies, and identify and counteract faulty plant system elements.

The TSC personnel shall have ready access to up-to-date records, operational specifications, and procedures that include but are not limited to:

- Plant technical specifications,
- Plant operating procedures,
- Emergency operating procedures,
- Final Safety Analysis Report,
- Plant operating records,
- Plant operations reactor safety committee records and reports,
- Records needed to perform the functions of the EOF when it is not operational,

and up-to-date, as-built drawings, schematics, and diagrams showing:

- Conditions of plant structures and systems down to the component level, and

- In-plant locations of these systems.

All of the above records shall be available in the TSC in current form when this facility is fully implemented. These records will be updated as necessary to ensure currency and completeness. The method of storage and presentation of the TSC records shall ensure ease of access under emergency conditions.

3. OPERATIONAL SUPPORT CENTER

3.1 Functions

The operational support center (OSC) is an onsite area separate from the control room and the TSC where licensee operations support personnel will assemble in an emergency. The OSC shall:

- Provide a location where plant logistic support can be coordinated during an emergency, and

- Restrict control room access to those support personnel specifically requested by the shift supervisor.

When the OSC is activated, it shall be supervised by licensee operations management personnel designated in the licensee's emergency plan to perform these functions.

3.2 Habitability

No specific habitability criteria are established for the OSC. If the OSC habitability is not comparable to that of the control room, the licensee's emergency plan shall include procedures for evacuation of OSC personnel in the event of a large radioactive release. These procedures also shall include provisions for the performance of the OSC functions by essential support personnel from other onsite locations.

3.3 Communications

The OSC shall have direct communications with the control room and with the TSC so that the personnel reporting to the OSC can be assigned to duties in support of emergency operations. The OSC communications system shall consist of one dedicated telephone extension to the control room, one dedicated telephone extension to the TSC, and one dial telephone capable of reaching onsite and offsite locations, as a minimum. Direct voice intercommunications and/or reliable direct radio communications may be used to supplement these telephone communication links.

4. EMERGENCY OPERATIONS FACILITY

4.1 Functions

The emergency operations facility (EOF) is a licensee controlled and operated offsite support center. The EOF will have facilities for:

- Management of overall licensee emergency response,
- Coordination of radiological and environmental assessment,
- Determination of recommended public protective actions, and
- Coordination of emergency response activities with Federal, State, and local agencies.

When the EOF is activated, it shall be staffed by licensee, Federal, State, local and other emergency personnel designated by the emergency plan to perform these functions. It shall be the location where the licensee provides overall management of licensee resources in response to an emergency having actual or potential environmental consequences. A designated senior licensee official will manage licensee activities in the EOF to support the designated official in the TSC and the senior reactor operator designated the shift supervisor in the control room.

Facilities shall be provided in the EOF for the acquisition, display, and evaluation of all radiological, meteorological, and plant system data pertinent to determine offsite protective measures. These facilities will be used to evaluate the magnitude and effects of actual or potential radioactive releases from the plant and to determine offsite dose projections. Facilities used in performing essential EOF functions must be located within the EOF complex. However, supplemental calculations and analytical support of EOF evaluations may be provided from facilities outside the EOF. The licensee also may use the EOF as the post-accident recovery management center.

The licensee shall use the EOF to coordinate its emergency response activities with those of local, State, and Federal agencies, including the NRC. Licensee personnel in the EOF will use the evaluations of offsite effects to make protective action recommendations for the public to State and local emergency response agencies.

State and local agencies will be responsible for implementing emergency response actions involving the general public. The State and local agencies may operate from the EOF or from their own control centers at other locations, dependent upon the site-specific provisions of the emergency plan at each plant. Collocation of offsite authorities at the EOF for the purpose of offsite dose estimation is encouraged.

At the licensee's option, the EOF may be a location for information dissemination to the public via the news media by designated spokespersons in accordance with the licensee's emergency plan. Provisions to allow periodic briefings of a press pool at the EOF should also be made. Actual use of this provision would depend on specific accident conditions and the emergency plan of the licensee.

The licensee shall provide normal industrial security for the EOF complex during normal operating conditions. This protection is required to ensure EOF activation readiness for an emergency by the exclusion of unauthorized persons.

After the EOF is activated, security protection shall be upgraded to restrict access to those personnel assigned to this facility. Any location provided in the EOF for news media briefings shall be outside of the controlled access area.

To maintain a proper level of readiness, the EOF shall be activated periodically for training and for emergency preparedness exercises as specified in the licensee's emergency plan. The EOF facility may be used by designated licensee personnel for normal daily operations as well as for training and exercises. Use of the EOF during normal operations shall be limited to activities that will not degrade EOF activation, operations, or reliability.

4.2 Location, Structure, and Habitability

The location of the EOF, and whether a backup facility is required, should consider the following factors:

- Whether the location provides optimum functional and availability characteristics for carrying out the licensee functions specified for the EOF (i.e., overall strategic direction of licensee onsite and support operations, determination of public protective actions to be recommended by the licensee to offsite officials, and coordination of the licensee with Federal, State, and local organizations).

- Whether the EOF functions would be interrupted during radiation releases for which it was necessary to recommend protective actions for the public to offsite officials.

It is strongly recommended that the EOF location be coordinated with State and local authorities to improve the relationship between the licensee and offsite organizations.

The habitability criteria for the EOF determined by its location are given in Table 2 below.

Table 2.   Relation of EOF Location to Habitability Criteria

	EOF Criteria	
Item Needed	Distance within 10 mi of the TSC	Distance at or beyond 10 mi of the TSC[1]
Structure	Well engineered for design life of plant[2]	Well engineered for design life of plant[2]
Protection factor[3]	≥ 5	None
Ventilation protection	Isolation with HEPA[4] filters (no charcoal)	None
Backup EOF[5]	Located within 10 to 20 mi of the TSC	None

[1] Specific Commission approval is required for EOF locations beyond 20 miles of the TSC. For these cases, provisions must include arrangements to locate the NRC staff closer to the reactor.

[2] As an example of "well engineered," refer to the Uniform Building Code. In addition, it must be able to withstand adverse conditions of high winds (other than tornadoes) and floods. Winds and floods with a 100-yr recurrence frequency are acceptable for a design basis.

[3] Protection factor is defined in terms of the attenuation of 0.7 MeV gamma radiation. As a minimum, the protection factor only applies to those areas of the EOF in which dose assessments, communications, and decisionmaking take place.

[4] Ventilation system shall function in a manner comparable to the control room and TSC systems; but need not be seismic Category I qualified, redundant, instrumented, or automatically activated.

[5] Need not be a separate, dedicated facility, but, when activated, shall provide continuity of dose prediction and decisionmaking functions by arranging for portable backup equipment. No special provisions for protection factors or ventilation protection are needed.

Licensees who cannot meet the requirements of size and habitability for the EOF must submit to NRC a request for an exception. This request must include justification for the exception and an alternate proposal. NRC will review requests for exceptions on a case-by-case basis.

18

4.3 Staffing and Training

The EOF shall be staffed to provide the overall management of licensee resources and the continuous evaluation and coordination of licensee activities during and after an accident. Upon EOF activation, designated personnel shall report directly to the EOF to achieve full functional operation within 1 hour. A senior management person designated by the licensee shall be in charge of all licensee activities in the EOF. The EOF staff will include personnel to manage the licensee onsite and offsite radiological monitoring, to perform radiological evaluations, and to interface with offsite officials. The EOF staff assignments shall be part of the licensee's emergency plan. The specific number and type of personnel assigned to the EOF may vary according to the emergency class. The staffing for each emergency class shall be fully detailed in the licensee's emergency plan. Operating procedures and staff training in the use of data systems and instrumentation shall contain guidance on the limitations of instrumentation including whether the information can be relied upon following serious accidents.

In order to function effectively, the EOF staff personnel must be aware of their responsibilities during an accident. To maintain proficiency, the EOF staff shall participate in EOF activation drills, which shall be conducted periodically in accordance with the licensee's emergency plan. These drills shall include operation of all facilities that will be used to perform the EOF functions, including any support facilities located outside the EOF.

4.4 Size

The EOF building or building complex shall be large enough to provide the following:

- Working space for the personnel assigned to the EOF as specified in the licensee's emergency plan, including State and local agency personnel, at the maximum level of occupancy without crowding (minimum size of working space provided shall be approximately 75 sq ft/person);

- Space for EOF data system equipment needed to transmit data to other locations;

- Sufficient space to perform repair, maintenance, and service of equipment, displays, and instrumentation;

- Space for ready access to communications equipment by all EOF personnel who need communications capabilities to perform their functions;

- Space for ready access to functional displays of EOF data;

- Space for storage of plant records and historical data or space for means to readily acquire and display those records; and

- Separate office space to accommodate at least five NRC personnel during periods that the EOF is activated for emergencies.

The EOF working space shall be sized for at least 35 persons, including 25 persons designated by the licensee, 9 persons from NRC, and 1 person from FEMA. This minimum size shall be increased if the maximum staffing levels specified in the licensee's emergency plan, including representatives from State and local agencies, exceed 25 persons.

4.5 Radiological Monitoring

To ensure adequate radiological protection of EOF personnel, radiation monitoring systems shall be provided in the EOF. These monitoring systems may be composed of installed monitors or dedicated, portable monitoring equipment. These systems shall continuously indicate radiation dose rates and airborne radioactivity concentrations inside the EOF while it is in use during an emergency. These monitoring systems shall include local alarms with trip levels set to provide early warning to EOF personnel of adverse conditions that may affect the habitability of the EOF. Detectors to distinguish the presence or absence of radioiodines at concentrations as low as 10^{-7} microcuries/cc shall be provided.

4.6 Communications

The EOF shall have reliable voice communications facilities to the TSC, the control room, NRC, and State and local emergency operations centers. The normal communication path between the EOF and the control room will be through the TSC. The primary functions of the EOF voice communications facilities will be:

- EOF management communications with the designated senior licensee manager in charge of the TSC,

- Communications to manage licensee emergency response resources,

- Communications to coordinate radiological monitoring,

- Communications to coordinate offsite emergency response activities, and

- Communications to disseminate information and recommended protective actions to responsible government agencies.

The EOF voice communications facilities shall include reliable primary and backup means of communication. Voice communications may include private telephones, commercial telephones, radio networks, and intercommunications systems as appropriate to accomplish the EOF functions during emergency conditions. Existing licensee communications systems may be used if the systems can be demonstrated to be reliable under emergency conditions and if they are adequate to meet the added EOF criteria. The licensee shall provide a means for EOF telephone access to commercial telephone common-carrier services that bypasses any local telephone switching facilities that may be susceptible to loss of power during emergencies. The licensee shall ensure that spare commercial telephone lines to the plant are available for use by the EOF during emergencies.

The EOF voice communications equipment shall include:

- Hotline telephone (located in the NRC office space) on the NRC emergency notification system (ENS) to the NRC Operations Center;

- Dedicated telephone (located in the NRC office space) on the NRC health physics network (HPN);

- Dedicated telephones for management communications with direct access to the TSC and the control room;

- Dial telephones reserved for EOF use to provide access to onsite and offsite locations;

- Intercommunications systems between work areas of the EOF, if needed, for EOF functional performance or if the EOF is comprised of separate functional areas or separate buildings;

- Radio communications to licensee mobile monitoring teams;

- Communications to State and local operations centers; and

- Communications to facilities outside the EOF used to provide supplemental support for EOF evaluations.

The EOF communication system shall also include designated telephones (in addition to the ENS and HPN telephones) for use by NRC personnel. The licensee shall provide at least three dial telephone lines for such NRC use while the EOF is activated. The licensee shall also furnish the access facilities and cables to the NRC for the ENS and HPN telephones.

Facsimile transmission capability between the EOF, the TSC, and the NRC Operations Center shall be provided.

4.7 Instrumentation, Data System Equipment, and Power Supplies

Equipment shall be provided to gather, store, and display data needed in the EOF to analyze and exchange information on plant conditions with the designated senior licensee manager in charge of the TSC. The EOF data system equipment shall perform these functions independently from actions in the control room and without degrading or interfering with control room and plant functions. EOF instrumentation, data system equipment, and power supplies need not meet safety grade or Class 1E criteria. When signals to the EOF are received from sensors providing signals to safety system equipment or displays, suitable isolation in accordance with GDC 22, 23, and 24 shall be provided to ensure that the EOF systems cannot degrade performance of the safety system equipment of displays.

The EOF electrical equipment load shall not degrade the capability or reliability of any safety-related power source. Circuit transients or power supply failures and fluctuations shall not cause a loss of any stored data vital to the EOF functions.

The total EOF data system shall be designed to achieve an operational unavailability goal of 0.01 during all plant operating conditions above cold shutdown as described in Section 1.5.

The plant data processor (plant process computer) may provide, if stringent conditions are met, the EOF data system functions or may have data for the EOF system transmitted through it. In either case, the performance of the plant data processor and its related instrumentation and equipment shall be included when determining the EOF data system unavailability. In addition, the computational capacity and data throughput of the plant processor must be sufficient to accommodate the combined computational and input-output loads of the EOF system and the other functions being performed by the plant processor. EOF data processed through the plant data processor shall be continuously available to the EOF with the same reliability and accuracy required of a data system designed to operate independent of the plant data processor. Use of the plant data processor shall not degrade the integrity of data supplied to the EOF or the security of the software used to process EOF data.

The SPDS display equipment used in the EOF need not be seismically qualified and shall meet only the EOF data system equipment reliability and performance criteria. This display is necessary for licensee corporate management and the NRC Regional Director to communicate more easily with the TSC or the control room and to have a better understanding of the plant status. The design of the EOF data system equipment shall incorporate human-factors engineering with consideration for both operating and maintenance personnel.

4.8 Technical Data and Data System

The EOF technical data system will receive, store, process and display information sufficient to perform assessments of the actual and potential onsite and offsite environmental consequences of an emergency condition. Data providing information on the general condition of the plant also shall be available for display in the EOF for utility resource management.

The EOF data set shall include radiological, meteorological, and other environmental data as needed to:

- Assess environmental conditions,
- Coordinate radiological monitoring activities, and
- Recommend implementation of offsite emergency plans.

As a minimum EOF data set, sensor data of the Type A, B, C, D, and E variables specified in Reg. Guide 1.97, Revision 2, and of those meteorological variables specified in proposed Revision 1 to Regulatory Guide 1.23, "Meteorological Measurements Programs in Support of Nuclear Power Plants," and in NUREG-0654, Revision 1, Appendix 2, shall be available for display in the EOF. All data that are available for display in the TSC, including data transmitted from the plant to NRC, shall be part of the EOF data set.

The accuracy of data in the EOF must be consistent with the data accuracy needed to perform the EOF functions. The accuracy of data displays in the EOF shall be equivalent to that for the data displayed in the TSC. The time resolution of data requisition shall be sufficient to provide data without loss of information during transient conditions. The time resolution required for each sensor signal shall depend on the potential transient behavior of the variable being measured. The EOF data displays of Reg. Guide 1.97 variables shall meet the criteria for EOF data but need not meet the design and qualification criteria in Reg. Guide 1.97 for display of those variables in the control room.

22

Data storage capability shall be provided for the EOF data set. At least 2 hours of pre-event data and 12 hours of post-event data shall be recorded. The sample frequency shall be chosen to be consistent with the use of the data. Capacity to record at least two weeks of additional post-event data with reduced time resolution shall be provided. Archival data storage and the capability to transfer data between active memory and archival data storage without interrupting EOF data acquisition and displays shall be provided for all EOF data. A sufficient number of data display devices shall be provided in the EOF to allow all EOF personnel to perform their assigned tasks with unhindered access to alphanumeric and/or graphical representations of:

- Plant systems variables,
- In-plant radiological variables,
- Meteorological information, and
- Offsite radiological information.

Trend-information display and time-history display capability is required in the EOF to give EOF personnel a dynamic view of plant systems, radiological status, and environmental status during an emergency. The EOF displays shall be designed so that callup, manipulation, and presentation of data can be easily performed. The displays shall be partitioned to facilitate the retrieval of information by the different functional groups in the EOF. This may be accomplished with either separate display units or by logically separated information display pages available on a callup basis at each data display unit. The EOF data display formats shall present information so that it can be easily understood by the EOF personnel operating the system. If display capabilities for news media briefings are provided in the EOF, these displays shall be separated physically from the EOF functional displays. Human-factors engineering shall be incorported in the design of the EOF.

The SPDS shall also be displayed in the EOF. This duplication will provide licensee management and NRC representatives information about the current reactor systems status and will facilitate communications among the control room, TSC, and EOF. If the SPDS system in the control room is composed of multiple display units, multiple displays must also be provided in the EOF.

4.9 Records Availability and Management

The EOF shall have ready access to up-to-date plant records, procedures, and emergency plans needed to exercise overall management of licensee emergency response resources. The EOF records shall include, but shall not be limited to:

- Plant technical specifications,
- Plant operating procedures,
- Emergency operating procedures,
- Final Safety Analysis Report,
- Up-to-date records related to licensee, State, and local emergency response plans,
- Offsite population distribution data,
- Evacuation plans,
- Environs radiological monitoring records, and
- Licensee employee radiation exposure histories.

and up-to-date drawings, schematics and diagrams showing:

- Conditions of plant structures and systems down to the component level and
- In-plant locations of these systems.

These records shall either be stored and maintained in the EOF (such as hard copy or microfiche) or shall be readily available via transmittal to the EOF from another records storage location. The method of storage and presentation of the EOF records shall ensure ease of access under emergency conditions. The records available to the EOF shall be completely updated as necessary to ensure currency and completeness.

5. SAFETY PARAMETER DISPLAY SYSTEM

5.1 Function

The purpose of the safety parameter display system (SPDS) is to assist control room personnel in evaluating the safety status of the plant. The SPDS is to provide a continuous indication of plant parameters or derived variables representative of the safety status of the plant. The primary function of the SPDS is to aid the operator in the rapid detection of abnormal operating conditions. The functional criteria for the SPDS presented in this section are applicable for use only in the control room.

It is recognized that, upon the detection of an abnormal plant status, it may be desirable to provide additional information to analyze and diagnose the cause of the abnormality, execute corrective actions, and monitor plant response as secondary SPDS functions.

As an operator aid, the SPDS serves to concentrate a minimum set of plant parameters from which the plant safety status can be assessed. The grouping of parameters is based on the function of enhancing the operator's capability to assess plant status in a timely manner without surveying the entire control room. However, the assessment based on SPDS is likely to be followed by confirmatory surveys of many non-SPDS control room indicators.

Human-factors engineering shall be incorporated in the various aspects of the SPDS design to enhance the functional effectiveness of control room personnel. The design of the primary or principal display format shall be as simple as possible, consistent with the required function, and shall include pattern and coding techniques to assist the operator's memory recall for the detection and recognition of unsafe operating conditions. The human-factored concentration of these signals shall aid the operator in functionally comparing signals in the assessment of safety status.

All data for display shall be validated where practicable on a realtime basis as part of the display to control room personnel. For example, redundant sensor data may be compared, the range of a parameter may be compared to predetermined limits, or other quantitive methods may be used to compare values. When an unsuccessful validation of data occurs, the SPDS shall contain means of identifying the impacted parameter(s). Operating procedures and operator training in the use of the SPDS shall contain information and provide guidance for the resolution of unsuccessful data validation. The objective is to ensure that the SPDS presents the most current and accurate status of the plant possible and is not compromised by unidentified faulty processing or failed sensors.

The SPDS shall be in operation during normal and abnormal operating conditions. The SPDS shall be capable of displaying pertinent information during steady-state and transient conditions. The SPDS shall be capable of presenting the magnitudes and the trends of parameters or derived variables as necessary to allow rapid assessment of the current plant status by control room personnel.

The parameter trending display shall contain recent and current magnitudes of the parameter as a function of time. The derivation and presentation of parameter trending during upset conditions is a task that may be automated, thus freeing the operator to interpret the trends rather than generate them. Display of time derivatives of the parameters in lieu of trends to both optimize operator-process communication and conserve space may be acceptable.

The SPDS may be a source of information to other systems, and the functional criteria of these systems shall state the required interfaces with the SPDS. Any interface between the SPDS and a safety system shall be isolated in accordance with the safety system criteria to preserve channel independence and ensure the integrity of the safety system in the case of SPDS malfunction. Design provisions shall be included in the interfaces between the SPDS and nonsafety systems to ensure the integrity of the SPDS upon failure of nonsafety equipment.

A qualification program shall be established to demonstrate SPDS conformance to the functional criteria of this document.

5.2 Location

The SPDS shall be located in the control room with additional SPDS displays provided in the TSC and the EOF. The SPDS may be physically separated from the normal control board; however, it shall be readily accessible and visible to the shift supervisor, control room senior reactor operator, shift technical advisor, and at least one reactor operator from the normal operating area. If the SPDS is part of the control board, it shall be easily recognizable and readable.

5.3 Size

The SPDS shall be of such size as to be compatible with the existing space in the control area. The SPDS display shall be readable from the emergency operating station of the control room senior reactor operator. It shall not interfere with normal movement or with full visual access to other control room operating systems and displays.

5.4 Staffing

The SPDS shall be of such design that no operating personnel in addition to the normal control room operating staff are required for its operation.

5.5 Display Considerations

The display shall be responsive to transient and accident sequences and shall be sufficient to indicate the status of the plant. For each mode of plant operation, a single primary display format designed according to acceptable human-factors principles (a limited number of parameters or derived variables and their trends in an organized display that can be readily interpreted by an operator) shall be displayed, from which plant safety status can be inferred. It is recognized that it may be desirable to have the capability to recall additional data on secondary formats or displays.

The primary display may be individual plant parameters or may be composed of a number of parameters or derived variables giving an overall system status. The basis for the selection of the minimum set of parameters in the primary display shall be documented as part of the design.

The important plant functions related to the primary display while the plant is generating power shall include, but not be limited to:

- Reactivity control
- Reactor core cooling and heat removal from primary system
- Reactor coolant system integrity
- Radioactivity control
- Containment integrity

The SPDS may consist of several display formats as appropriate to monitor and present the various parameters or derived variables. For each plant operating mode, these formats may either be automatically displayed or manually selected by the operator to keep control room operating personnel informed. Flexibility to allow for interaction by the operator is desirable in the display designs. Also, where feasible, the SPDS should include some audible notification to alert personnel of an unsafe operating condition.

The SPDS need not be limited to the previously stated functions. It may include other functions that aid operating personnel in evaluating plant status. It is desirable that the SPDS be sufficiently flexible to allow for future incorporation of advanced diagnostic concepts and evaluation techniques and systems.

5.6 Design Criteria

The total SPDS need not be Class 1E or meet the single-failure criterion. The sensors and signal conditioners (such as preamplifiers, isolation devices, etc.) shall be designed and qualified to meet Class 1E standards for those SPDS parameters that are also used by safety systems. Furthermore, sensors and signal conditioners for those parameters of the SPDS identical to the parameters specified within Reg. Guide 1.97 shall be designed and qualified to the criteria stated in Reg. Guide 1.97. For SPDS application, it is also acceptable to have Class 1E qualified devices from the sensor to a post-accident-accessible location, such as outside containment, and then non-1E devices from containment to the display (or processor) on the presumption that these components can be repaired or replaced in an accident environment. The processing and display devices of the SPDS shall be of proven high quality and reliability.

The function of the SPDS is to aid the operator in the interpretation of transients and accidents. This function shall be provided during and following all events expected to occur during the life of the plant, including earthquakes. To achieve this function, the display system shall not only take adequate account of human factors--the man-machine interface--but shall also be sufficiently durable to function during and after earthquakes. Because of current technology, it may not be possible to satisfy these criteria within one SPDS system.

From an operational viewpoint, it is preferred that only one display system be used for evaluating the safety status of the plant. One display system simplifies the man-machine interface and thus minimizes operator errors. However, in recognition of the restraints imposed by current technology, an alternative is to design the overall SPDS function with a primary and a backup display system: (1) the primary SPDS display would have high performance and flexibility and be human factored but need not be seismically qualified; and (2) the backup display system would be operable during and following earthquakes, such as the normal control room displays needed to comply with Reg. Guide 1.97. The display system (or systems) provided for the SPDS function shall be capable of functioning during and following all design basis events for the plant.

In all cases, both the primary SPDS display and the backup SPDS seismically qualified portion of the display shall be sufficiently human factored in its design to allow the control room operations staff to perform the safety status assessment task in a timely manner. Dependence on poorly human-engineered Class 1E seismically qualified instruments that are scattered over the control board, rather than concentrated for rapid safety status assessment, is not acceptable for this function. An acceptable approach would be to concentrate the seismically qualified display into one segment of the control board.

The dynamic loading limitations of the SPDS design shall be defined and incorporated into the training program. The control room operations staff shall be provided with sufficient information and criteria to allow for performance of an operability evaluation of SPDS if an earthquake should occur.

The SPDS as used in the control room shall be designed to an operational unavailability goal of 0.01, as defined in Section 1.5 of this document. The cold shutdown unavailability goal for the SPDS during the cold shutdown and refueling modes for the reactor shall be 0.2, as defined in Section 1.5.

Technical specifications shall be established to be consistent with the unavailability design goal of the SPDS and with the compensatory measures provided during periods when the SPDS is inoperable. Operation of the plant with the SPDS out of service is allowed provided that the control board is sufficiently human factored to allow the operations staff to perform the safety status assessment task in a timely manner. Dependence on poorly human-engineered instruments that are scattered over the control board rather than concentrated for rapid safety status assessment is not acceptable for this function.

6. NUCLEAR DATA LINK

6.1 Function

When a significant incident occurs at a nuclear power plant, the NRC will activate the Executive Team (ET) at the NRC Operations Center to oversee the agency response. The ET for reactor events consists of the Director of the NRC response (Chairman or designated alternate), the Executive Director for Operations, the Director of the Office of Inspection and Enforcement, and the Director of the Office of Nuclear Reactor Regulation. When the ET is activated, the Regional Director and Regional support staff leave immediately for the affected site, and a headquarters technical support group is called in to provide assistance to the ET and to NRC site personnel.

The primary role of each of these components of the NRC Incident Response Program is to monitor the event, independently evaluate the situation, provide advice and assistance to the licensee and offsite authorities, and inform officials and the general public about the radiological conditions on site and around the facility and the physical condition of facility.

Until the Regional Director (or other designated senior NRC official) arrives at the site, the NRC Operations Center must carry out the roles described above. When the Regional Director or other designated senior NRC official arrives at the site, it is contemplated that responsibility for managing the NRC operations in and around the plant will be transferred to that individual at the site (i.e., NRC Director of Site Operations).

After the responsibility for managing the NRC site operations has been transferred to the Director of Site Operations, the headquarters technical support groups will provide direct support to him by continuing to evaluate information provided by site personnel. In addition, certain key decisions, particularly those relating to recommendations for actions affecting the general public and those involving changes in the NRC's role in responding to the accident, may remain with the Director of the NRC response. The NRC role will not extend to any manipulation of nuclear power plant controls nor will it normally include directing licensee actions. However, in extreme cases, NRC may direct that certain operations be performed by the licensee at the nuclear plant. In such an unlikely situation, it is expected that any direction from NRC would be provided to licensee management from the Director of Site Operations after his arrival onsite or from the Director of NRC response (Chairman or designated alternate) prior to that time. (A more detailed description of the manner in which NRC will respond to significant incidents can be found in NUREG-0728, "Report to Congress: NRC Incident Response Plan," and in NUREG-0730, "Report to Congress on the Acquisition of Reactor Data for the NRC Operations Center".) To fulfill these functions of providing support, NRC must have reliable and timely data from the nuclear power plant. This will be obtained through a nuclear data link (NDL), a data transmission system designed to send a specified set of variables from the plant to the NRC Operations Center.

6.2 Description

The nuclear data link (NDL) processes and transmits certain reactor process variables, and radiological and site meteorological data from each operating nuclear power plant to the NRC Operations Center. The NDL system is comprised of a data acquisition system, an NDL terminal (both of which are located on site), and an Operations Center system at NRC headquarters. Figure 1 shows a block diagram of how the NDL could be connected to the emergency response data acquisition system and is explained in the following paragraphs.

28

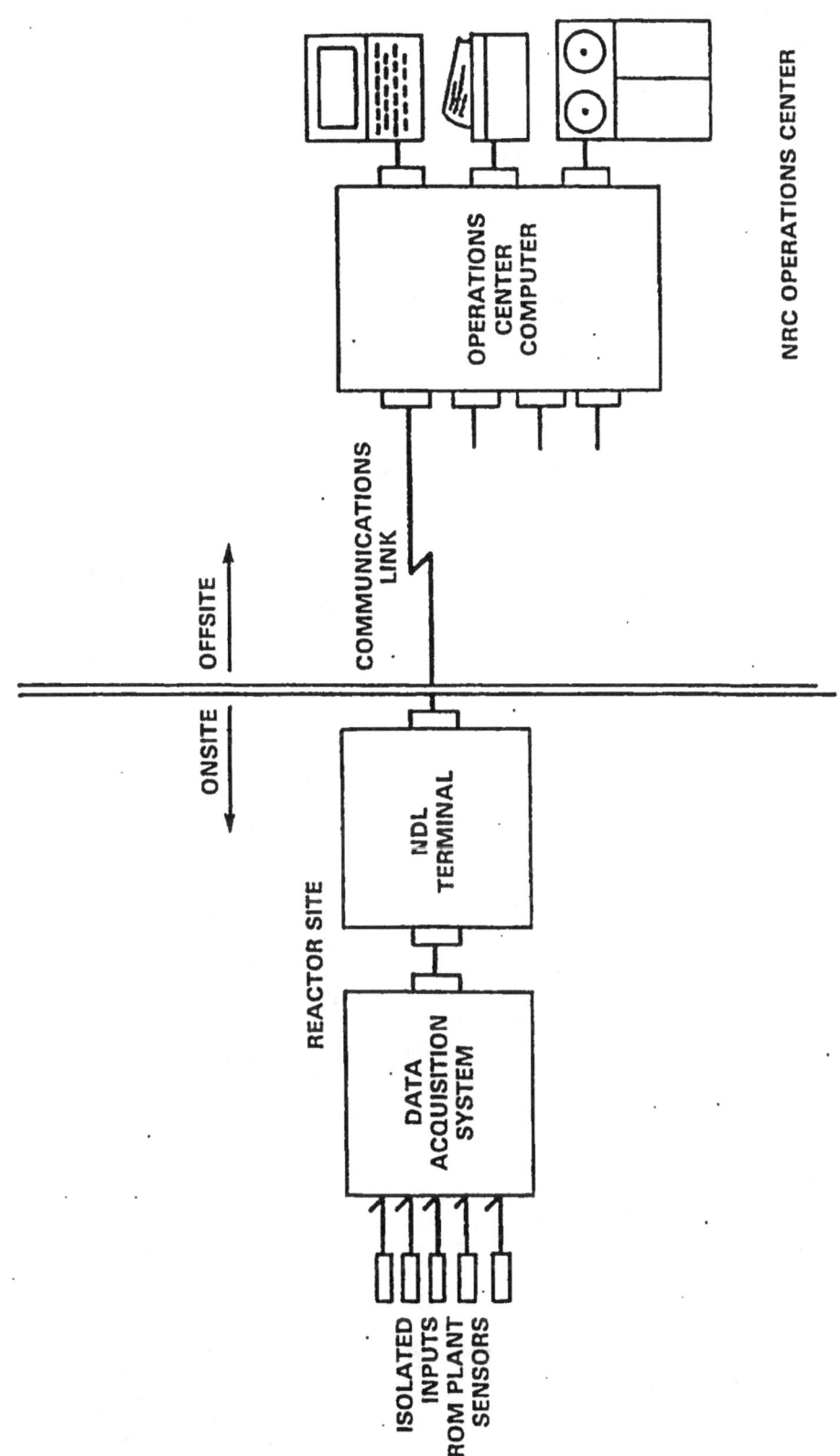

Figure 1. NDL System Overview

29

The data acquisition system (DAS) acquires data for the variables listed in Reg. Guide 1.97, Rev. 2, and Reg. Guide 1.23 and also may include other site-specific variables for transmission to the TSC, EOF, and SPDS as determined by the licensee. The DAS also has Reg. Guide 1.97 data available to be transmitted to the NRC Operations Center over the NDL and to vendors and State or local authorities over other data links. Reg. Guide 1.97 information currently lists approximately 60 variables, depending on whether a PWR or BWR facility is being considered. The DAS performs multiplexing of incoming signals and conversion of raw data into engineering units. The data are scaled and time tagged appropriately and converted into a standardized, digital format for transmission.

Sensor sampling by the DAS and data transmission to the NDL terminal must occur at intervals of not greater than one minute. The analog-to-digital conversion performed on the input data transmitted shall have 12-bit resolution. Each reading shall be time tagged.

A data-access approach discussed in Appendix E of NUREG/CR-1451 may be used for transients. With that approach, the peak value and time integral of the transient are captured, digitized, formatted, and transmitted at one-minute intervals.

An NDL terminal at the plant may accept data from the DAS and transmit it to the NRC Operations Center. The terminal would handle the communications protocol as well as error detection and correction. The NDL terminal hardware and software would be a standardized unit specified by the NRC. Specification for this unit would be developed as a part of the NDL system design.

The Operations Center system makes data obtained from the licensee facility available for viewing in the NRC Operations Center. Facilities for retention of data will exist at the NRC Operations Center. The Operations Center subsystem, as presently envisioned, will include a general-purpose computer that is capable of receiving data from any plant. Video data terminals, printers, magnetic memory storage, and miscellaneous peripherals will comprise the balance of the equipment at the NRC Operations Center. The peripherals provide the man-machine interface to access and display numerical and graphical representations of parametric data and provide data storage capabilities. The computer may be used to maintain a file of current data from each reactor site. When an incident occurs, long-term data retrieval and storage begins in the NRC Operations Center. Stored data from at least a 2-week period shall be available for recall and display, as well as for analysis and verification by NRC personnel.

6.3 NDL Interface

The interface between the data acquisition system and the NDL will require the use of a common communications protocol, to be defined later.

6.4 Environment

The DAS and the NDL interface shall meet the same environmental specifications that apply to the onsite technical support center. Power supply for the acquisition system and formatting equipment shall be high-reliability non-Class 1E power that is backed by battery power to eliminate momentary interruptions.

The licensee is responsible for the correction of any failure that results in degraded plant data being delivered to the NDL communications link at the specified intervals.

7. ACQUISITION AND CONTROL OF TECHNICAL DATA

7.1 Sources of Technical Data

Parameters specified in Reg. Guide 1.97, Rev. 2, and Reg. Guide 1.23 shall be provided to the data acquisition system (DAS). Isolation devices are needed for all signal interfaces with safety systems to prevent interference, degradation, or damage to any element of the safety system as specified in General Design Criterion 24, "Separation of Protection and Control Systems," and IEEE Standard 279-1971, Section 4.7, "Control and Protection System Interaction." The signals may be provided at a control room interface or input connection to the process control computer. These inputs shall not be processed by a software-programmable device, or any device controlled indirectly by software, before entering the DAS except for data received in accordance with Reg. Guide 1.23.

7.2 Acquisition of Data

Examples of data acquisition and distribution systems are shown in Figures 2 and 3 of this document. The configuration in Figure 3 is the anticipated data acquisition system that will be implemented. If the conditions of Section 2.8 are met, the configuration shown in Figure 2 could be acceptable. All inputs that interface with safety system signals must be isolated by an isolation device prior to connection to the DAS.

7.3 DAS Functional Limitations

The DAS must not be subject to external demands for processing or services that could degrade the needed reliability under accident conditions and must not be interrupted, delayed, or in any way impeded or degraded in its function by any such external demands or software installation or changes in any plant equipment. The only exception is the system's internal calibration and self-diagnostic routines. Output data from the DAS must be consistent with readings observed by the operators in the control room. To achieve this objective, verification and validation tests shall be performed to assure correlation of data obtained from each source.

7.4 DAS Design, Verification, and Configuration Control

Because the data acquisition system may be the basic source of data for all of the emergency facilities, its hardware and software configurations and changes shall be verified for reliability. Tests to demonstrate and evaluate the integrity of software and the integrated system are needed. These tests should be performed with the system operating continuously on live input signals in addition to satisfactory performance of static and dynamic test cases.

31

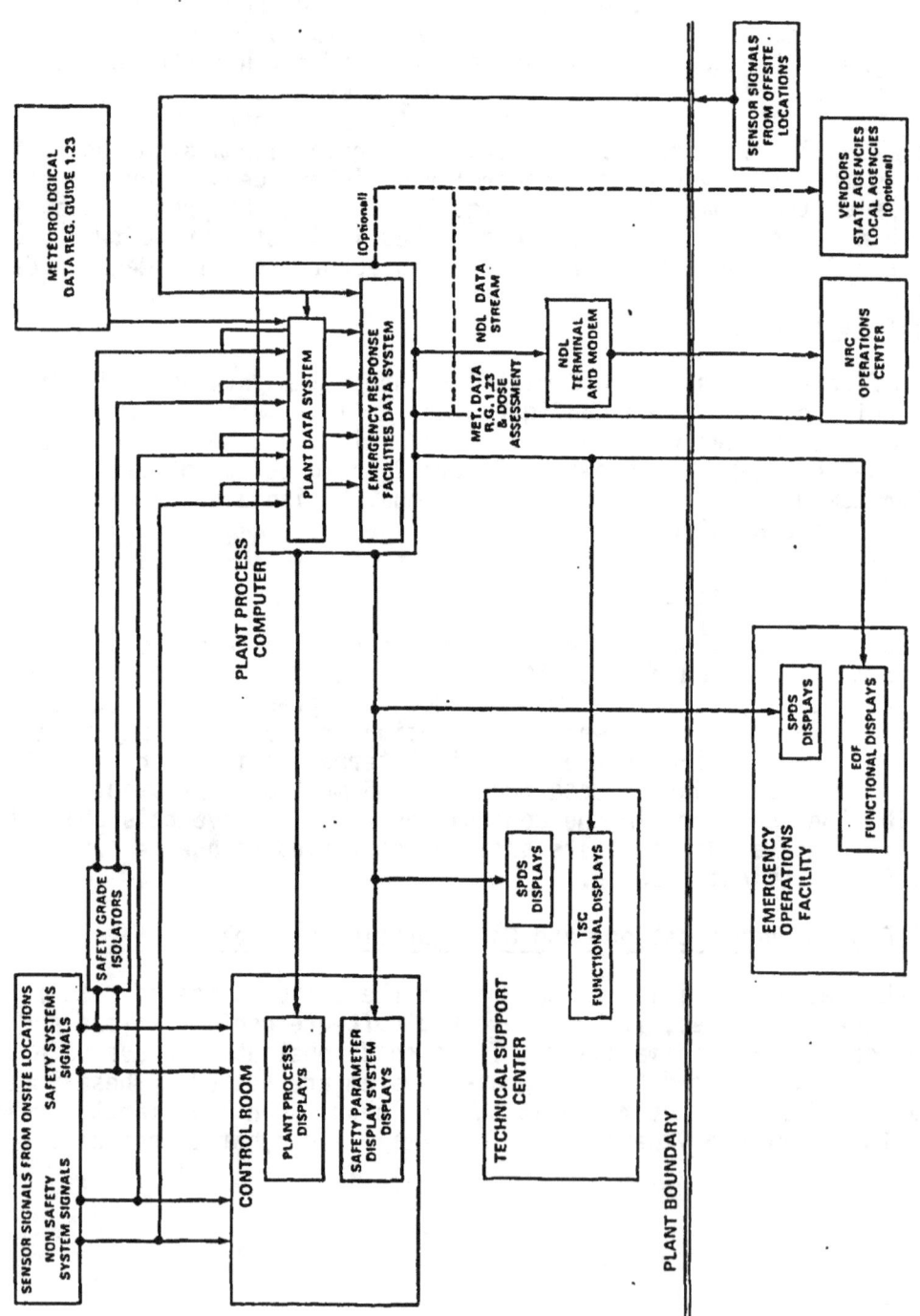

Figure 2. Example of a Functional Block Diagram of Data Flow Using Plant Process Computer.

32

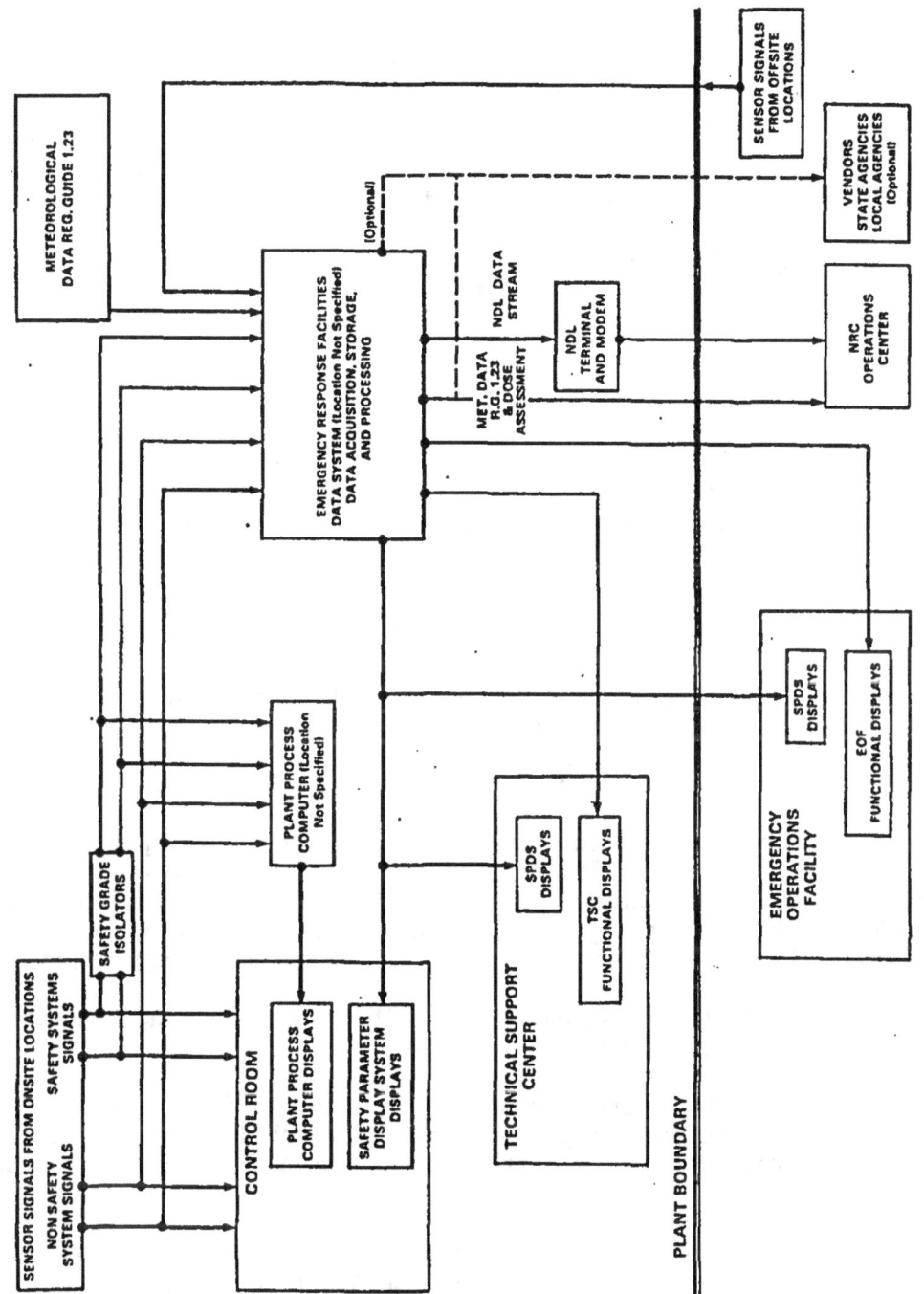

Figure 3. Example of Functional Block Diagram Flow Without Using Plant Process Computer.

33

The following are the primary criteria for qualification of changes in the DAS:

(1) The original development and subsequent changes are to be performed in accordance with documented quality assurance procedures that are to be available for review by the NRC staff.

(2) The original software and subsequent modifications are to be subjected to static and dynamic test programs to verify performance. Once the system becomes operational, all subsequent software changes and modifications must undergo static and dynamic tests to verify performance. After the performance has been verified, the software may be transferred to the DAS in accordance with an approved DAS update procedure.

These types of criteria were used to ensure software reliability in Position 19 of NUREG-0308, "Safety Evaluation Report Related to Operation of Arkansas Nuclear One, Unit 2," Rev. 1 and supplements.

The criteria for qualification of the DAS describe the changes on DAS operation and its effect on compatibility with the NDL. Records shall be kept of the date a particular change was implemented.

7.5 DAS Reliability

The data acquisition system shall provide data access for the SPDS, TSC, and EOF, as well as the NDL. Therefore, high DAS reliability is paramount to achieve accurate and dependable information on a continued basis. A moderate amount of excess capacity and the capability to expand modularly should be included to allow modest increases in parameter monitoring without serious impact on the system.

8. EMERGENCY RESPONSE FACILITY INTEGRATION

During emergency conditions, it is essential that there be a continuous high level of interaction and communications among key personnel in the control room, emergency response facilities, and NRC to ensure that all emergency actions are fully understood and coordinated.

The foregoing emergency response facilities (ERFs) are developed to function as an integrated system. The licensee's ERFs shall be designed to provide coordinated support to the control room during emergency operating conditions. These facilities shall be integrated into the licensee's emergency plan at each site to facilitate coordination with State and local emergency response facilities.

The systems design of the ERFs shall ensure that the following functional criteria of each facility are satisfied:

- The operation of any system or subsystem within the licensee's ERFs shall not degrade the performance or reliability of any reactor safety or control system or of any safety-related displays in the control room.

- Actions in the control room and operation of control room systems shall not degrade or interfere with the functional operation of ERF systems.

Normal operation of any system or subsystem in the ERFs shall not degrade or interfere with the functional operation of other systems in those facilities.

The DAS hardware and software shall be protected against unauthorized manipulation of or interference with input signals, data processing, data storage, and data output.

The ERF data system may be a fully or partially integrated data-processing system serving all emergency response facilities and systems. Alternatively, the ERF data system may be comprised of multiple data-processing units serving individual parts of the ERFs or providing separate functions.

Regulatory Guide 1.97, Revision 2, establishes criteria for accident-monitoring instrumentation to be displayed in the control room. The minimum data set that shall be available for display and use in the TSC and EOF shall include the following:

Type A, B, C, D, and E variables specified in Reg. Guide 1.97;

Meteorological variables specified in Reg. Guide 1.23 and NUREG-0654, Revision 1, Appendix 2; and

Variables displayed by the SPDS.

The acquisition and transmission of data of Reg. Guide 1.97 variables to the TSC and EOF need not meet the Reg. Guide 1.97 design and qualification criteria for display of that data in the control room.

Signals for ERF systems may be received from sensors providing signals to safety related systems. Suitable isolation devices must be used to ensure that the ERF systems cannot degrade performance of the reactor safety systems. The ERF data systems need not be safety-grade systems beyond the device providing isolation from safety-related systems.

9.   VERIFICATION AND VALIDATION CRITERIA

The design, development, qualification, and installation of the SPDS, TSC, EOF, and NDL facilities and systems shall be independently verified and validated by qualified personnel other than the original designers and developers. The independent verification and validation of these facilities are needed to provide assurance that highly reliable and available systems will be implemented for all plant facilities. The licensee's quality assurance organization may be used for the verification and validation program if it meets the independence criteria and is technically capable of performing these functions.

Because the SPDS, TSC, and EOF are important elements in the capability of a nuclear power plant to respond to an emergency, limiting conditions for operation in the plant technical specifications shall be established to specify compensating actions to be taken by the licensee when the SPDS, TSC, and EOF are not operational. Reportable outage periods under the technical specification shall include a report on compensatory measures, such as additional staffing, taken during the time the SPDS, TSC, EOF, or NDL is not operational.

The plant technical specifications shall include surveillance requirements to determine the operability of the ERFs. The SPDS, TSC, OSC, EOF, and NDL facilities shall be designed to provide for periodic testing to diagnose and recognize component degradation and systems malfunction.

10. BIBLIOGRAPHY

1. Letter from H. R. Denton, NRC, to All Operating Nuclear Power Plants, Subject: Discussions of Lessons Learned Short-Term Requirements, dated October 30, 1979.*

2. Letter from D. G. Eisenhut, NRC, to All Operating Nuclear Power Plants, Subject: NRC Nuclear Data Link (NDL), dated March 12, 1980.*

3. Letter from D. G. Eisenhut, NRC, to All Power Reactor Licensees, Subject: Clarification of NRC Site Requirements for Emergency Response Facilities at Each Site, dated April 25, 1980.*

4. U.S. Nuclear Regulatory Commission, "Safety Evaluation Report Related to Operation of Arkansas Nuclear One, Unit 2," USNRC Report NUREG-0308, November 1977, Supplement 1, March 1978. (Available from NTIS only.)

5. U.S. Nuclear Regulatory Commission, "TMI-2 Lessons Learned Task Force Status Report and Short-Term Recommendations," USNRC Report NUREG-0578, July 1979.**

6. U.S. Nuclear Regulatory Commission, "TMI-2 Lessons Learned Task Force Final Report," USNRC Report NUREG-0585, August 1979.**

7. U.S. Nuclear Regulatory Commission, "Criteria for Preparation and Evaluation of Radiological Emergency Response Plans and Preparedness in Support of Nulcear Power Plants," USNRC Report NUREG-0654, Rev. 1, November 1980.***

8. U.S. Nuclear Regulatory Commission, "NRC Action Plan Developed as a Result of the TMI-2 Accident," USNRC Report NUREG-0660, Vols. 1 and 2, May 1980.***

9. U.S. Nuclear Regulatory Commission, "Report to Congress: NRC Incident Response Plan," USNRC Report NUREG-0728, September 1980.**

10. U.S. Nuclear Regulatory Commission, "Report to Congress on the Acquisition of Reactor Data for the NRC Operations Center," USNRC Report NUREG-0730, September 1980.**

11. U.S. Nuclear Regulatory Commission, "Clarification of TMI Action Plan Requirements," USNRC Report NUREG-0737, November 1980.***

12. U.S. Nuclear Regulatory Commission, "Standard Review Plan for the Review of Safety Analysis Reports for Nuclear Power Plants - LWR Edition," USNRC Report NUREG-75/087, Section 6.4, "Habitability Systems," Rev. 1. (Available from NTIS only.)

13. U.S. Nuclear Regulatory Commission, "Conceptual and Programmatic Framework for the Proposed Nuclear Data Link," prepared by Sandia National Laboratories, USNRC Report NUREG/CR-1451, April 1980.**

14. U.S. Nuclear Regulatory Commission, Regulatory Guide 1.23, "Meteorological Measurement Programs in Support of Nuclear Power Plants," proposed Revision 1.†

15. U.S. Nuclear Regulatory Commission, Regulatory Guide 1.97, "Instrumentation for Light-Water-Cooled-Nuclear Power Plants to Assess Plant and Environs Conditions During and Following an Accident," Revision 2.†

16. 10 CFR Part 50, General Design Criterion 19, "Control Room."*

17. 10 CFR Part 50, General Design Criterion 24, "Separation of Protection and Control Systems."*

---

*Available in the NRC Public Document Room for inspection and copying for a fee.

**Available for purchase from the NRC/GPO Sales Program, U.S. Nuclear Regulatory Commission, Washington, DC 20555, and the National Technical Information Service, Springfield, VA 22161.

***Available free upon written request to the Division of Technical Information and Document Control, U.S. Nuclear Regulatory Commission, Washington, DC 20555.

†Available from the NRC/GPO Sales Program, U.S. Nuclear Regulatory Commission, Washington, DC 20555, Attention: Regulatory Guide Account.

NRC FORM 335
(7-77)

U.S. NUCLEAR REGULATORY COMMISSION

BIBLIOGRAPHIC DATA SHEET

1. REPORT NUMBER (Assigned by DDC)

NUREG-0696

4. TITLE AND SUBTITLE (Add Volume No., if appropriate)

Functional Criteria for Emergency Response Facilities

2. (Leave blank)

3. RECIPIENT'S ACCESSION NO.

7. AUTHOR(S)

Safety Data Integration Group

5. DATE REPORT COMPLETED

MONTH	YEAR
February	1981

9. PERFORMING ORGANIZATION NAME AND MAILING ADDRESS (Include Zip Code)

Safety Data Integration Group
U. S. Nuclear Regulatory Commission
Washingotn, D. C. 20555

DATE REPORT ISSUED

MONTH	YEAR
February	1981

6. (Leave blank)

8. (Leave blank)

12. SPONSORING ORGANIZATION NAME AND MAILING ADDRESS (Include Zip Code)

Office of Inspection and Enforcement
U. S. Nuclear Regulatory Commission
Washington, D. C. 20555

10. PROJECT/TASK/WORK UNIT NO.

11. CONTRACT NO.

13. TYPE OF REPORT

Technical Report

PERIOD COVERED (Inclusive dates)

15. SUPPLEMENTARY NOTES

14. (Leave blank)

16. ABSTRACT (200 words or less)

This report describes the facilities and systems to be used by nuclear power plant licensees to improve responses to emergency situations. The facilities include the Technical Support Center (TSC), Onsite Operational Support Center (OSC), and Nearsite Emergency Operations Facility (EOF), as well as a brief discussion of the emergency response function of the control room. The data systems described are the Safety Parameter Display System (SPDS) and Nuclear Data Link (NDL). Together, these facilities and systems make up the total Emergency Response Facilities (ERFs). Licensees should follow the guidance provided both in this report and in NUREG-0654 (FEMA-REP-1), Revision 1, for design and implementation of the ERFs.

17. KEY WORDS AND DOCUMENT ANALYSIS

17a. DESCRIPTORS

17b. IDENTIFIERS/OPEN-ENDED TERMS

18. AVAILABILITY STATEMENT

Unlimited

19. SECURITY CLASS (This report)
Unclassified

20. SECURITY CLASS (This page)
Unclassified

21. NO. OF PAGES

22. PRICE
$

www.ingramcontent.com/pod-product-compliance
Lightning Source LLC
Chambersburg PA
CBHW081236170526
45165CB00009B/3067